SPACE
Quiz
Book

Published by Collins
An imprint of HarperCollins Publishers
Westerhill Road
Bishopbriggs
Glasgow G64 2QT

www.harpercollins.co.uk

HarperCollinsPublishers, 1st Floor, Watermarque
Building, Ringsend Road, Dublin 4, Ireland

In association with National Geographic Partners, LLC

NATIONAL GEOGRAPHIC and the Yellow Border Design are trademarks
of the National Geographic Society and used under license.

First published 2021

ISBN 978-0-00-840936-4

10 9 8 7 6 5 4 3 2

A catalogue record for this book is available from the British Library.

Printed in India

If you would like to comment on any aspect of this book, please contact us
at the above address or online.
natgeokidsbooks.co.uk
collins.reference@harpercollins.co.uk

Paper from responsible sources.

Acknowledgements
Text by Richard Happer

Images
All images © Shutterstock.com

SPACE
Quiz
Book

300
brain busting
trivia questions

Contents

THE SOLAR SYSTEM

Arrange the planets of our Solar System in order of their position relative to the Sun.

Planet	Position relative to the Sun
Jupiter	1st planet from the Sun
Earth	2nd planet from the Sun
Neptune	3rd planet from the Sun
Uranus	4th planet from the Sun
Mercury	5th planet from the Sun
Mars	6th planet from the Sun
Venus	7th planet from the Sun
Saturn	8th planet from the Sun

Now arrange the planets of our Solar System in order of size.

Planet	Size
Jupiter	1 (smallest)
Earth	2
Neptune	3
Uranus	4
Mercury	5
Mars	6
Venus	7
Saturn	8 (biggest)

THE SOLAR SYSTEM

Planet	Position relative to the Sun
Jupiter	5th planet from the Sun
Earth	3rd planet from the Sun
Neptune	8th planet from the Sun
Uranus	7th planet from the Sun
Mercury	1st planet from the Sun
Mars	4th planet from the Sun
Venus	2nd planet from the Sun
Saturn	6th planet from the Sun

Planet	Size
Jupiter	8 (biggest)
Earth	4
Neptune	5
Uranus	6
Mercury	1 (smallest)
Mars	2
Venus	3
Saturn	7

THE SUN

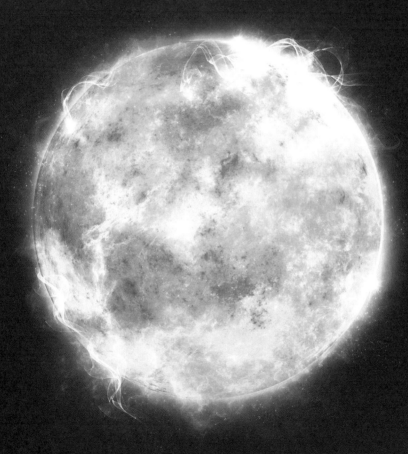

1 The Sun is the largest object in the Solar System. True or false?

2 How many Earths could you fit into the Sun?
a. 500,000   b. 1.3 million   c. 1.3 billion

3 What is the Sun made of?
a. Gases   b. Rock   c. Fire

4 How long has the Sun been burning?
a. 46 million years   b. 460 million years   c. 4.6 billion years

5 The Sun is not moving through space. True or false?

Week	
Mon	_____
Tue	_____
Wed	_____
Thu	_____
Fri	_____
Sat	_____
Sun	_____

6 Which day of the week is named after the Sun?
a. Sunday   b. Friday   c. Monday

7 How long does it take for light to travel from the Sun to the Earth?
a. 8 years and 20 days
b. 8 days and 20 minutes
c. 8 minutes and 20 seconds

8 What is the name for darker areas on the surface of the Sun caused by magnetic activity?
a. Freckles   b. Sunspots   c. Solar flares

9 Many human civilisations have worshipped the Sun. Which of these is NOT a sun god?
a. Selene   b. Ra   c. Apollo

10 You should never look directly at the Sun. True or false?

DID YOU KNOW?

Even though we use the colour yellow to represent the Sun, the Sun actually consists of all the colours mixed together, which then appears white to human eyes.

11

THE SUN

1 **True.** The Sun is gigantic. It weighs 700 times more than all the 8 planets of the Solar System combined.

2 **b.** 1.3 million – the Sun is huge!

3 **a.** The Sun is a ball of hot gases, mostly hydrogen.

4 **c.** The Sun has been burning for 4.6 billion years, but it is only half-way through its life.

5 **False.** The Sun and the whole Solar System are orbiting the centre of the Milky Way galaxy.

DID YOU KNOW?
The aurora borealis – also known as the Northern Lights – are the result of the Sun's solar winds interacting with Earth's atmosphere.

6 **a.** Sunday.

7 **c.** Light only takes 8 minutes and 20 seconds to reach Earth from the Sun, even though it is 150 million km away.

8 **b.** Sunspots can be over 10 times the diameter of the Earth.

9 **a.** Selene was the Greek goddess of the Moon.

10 **True.** Although the Sun lights our days and brings us warmth, it is very dangerous to look directly at it.

MERCURY

1

Who or what is Mercury named after?

a. Claudius Mercury, a famous ancient astronomer

b. The Roman messenger of the gods

c. Freddie Mercury, the singer

2

How long is a year on Mercury?

a. 365 Earth days

b. 200 Earth days

c. 88 Earth days

3

Because it is the closest planet to the Sun, the temperature on Mercury is hot all the time.

True or false?

4

How were the craters on the surface of Mercury created?

a. By underground caves collapsing

b. By rockets crash-landing on the planet

c. By asteroids and comets hitting the ground

5

How many spacecraft have visited Mercury?

a. 0   b. 2   c. 12

6

Which of these is NOT the name of a crater on Mercury?

a. J.K. Rowling Crater

b. Tolkien Crater

c. Disney Crater

7

How many moons does Mercury have?

a. 17   b. 0   c. 3

8

How long is a day on Mercury?

a. 10 Earth hours

b. 72 Earth hours

c. 59 Earth days

9

How big is Mercury?

a. A little bit bigger than our Moon

b. Half the size of our Moon

c. 10 times the size of our Moon

10

Mercury can be seen from Earth when it passes directly between us and the Sun. What is this event called?

a. Journey of Mercury

b. Shuffling of Mercury

c. Transit of Mercury

MERCURY

1

b. Mercury had a winged helmet and shoes so that he could travel quickly, just as the planet Mercury travels quickly around the Sun.

2

c. Mercury is the nearest planet to the Sun and the fastest moving, so it doesn't take long to make a complete orbit.

3

False. Mercury has no atmosphere to keep its heat, so it swings from being extremely hot during the day to extremely cold at night.

4

c. Mercury was bombarded by comets and asteroids from all directions.

5

b. *Mariner 10* flew by in 1974–5. *MESSENGER* orbited Mercury over 4000 times from 2011–15.

MARINER 10 ★ VENUS/MERCUR

US 10

Walt Disney at Madame Tussauds in Florida

6

a. Craters on Mercury are generally named after writers, artists and musicians (like Walt Disney, left – a writer, animator and producer) who were famous for at least 50 years and have been dead for at least 3 years.

7

b. Mercury and Venus are the only planets in the Solar System without moons.

8

c. Mercury may travel quickly round the Sun but it rotates relatively slowly on its axis.

9

a. Mercury is 4900 km in diameter and our Moon is 3476 km in diameter.

10

c. The next transit of Mercury will occur on 13 November 2032.

VENUS

1
Who is Venus named after?

a. The Norse god of fire

b. The Roman goddess of love

c. Venus Williams, the famous tennis player

2
How hot is the surface of Venus? It's hot enough to...

a. Fry an egg

b. Boil water

c. Melt metal

3
What is the best time to see Venus?

a. Just before sunrise or just after sunset

b. Midnight

c. During a lunar eclipse

4
What falls from the clouds on Venus?

a. Water

b. Sulphuric acid rain

c. Meatballs

5

How big is Venus?

a. Half the size of Earth

b. Almost the same size as Earth

c. Four times as big as Earth

6

Which of these features does Venus have?

a. A mountain taller than Mount Everest

b. Over one hundred large volcanoes

c. Violent thunderstorms with lightning and strong winds

7

Venus once had water oceans. What happened to them?

a. They evaporated away.

b. They were knocked into space by a massive asteroid impact.

c. They were boiled away by a volcanic eruption.

8

How long is a day on Venus?

a. 243 Earth hours

b. 243 Earth years

c. 243 Earth days

9

Why is Venus unique among the planets in the Solar System?

a. It is hollow.

b. It rotates in the opposite direction.

c. It doesn't orbit the Sun.

10

On Venus, a day lasts longer than a year.

True or false?

VENUS

1

b. Venus was named after the goddess of love because of its brightness and beauty. Venus is the brightest object in the night sky after the Moon.

2

c. The average temperature on Venus is around 470°C, which could melt some metals such as lead, tin and zinc.

3

a. Venus is known as the Morning Star and the Evening Star.

4

b. The thick sulphuric acid clouds on Venus would make the Sun look like a dull orange smudge.

5

b. Venus is 12,100 km in diameter; Earth is 12,700 km in diameter.

6

a, b and c. Venus has many unusual features.

7

a. Venus's atmosphere had so much greenhouse gas (gases that retain heat) that it couldn't cool down. It got hotter and hotter until its oceans boiled away.

8

c. Venus takes longer to rotate about its axis than any other planet in the Solar System.

9

b. Venus's spin on its axis is 'retrograde'. Only on Venus does the Sun rise in the west and set in the east.

10

True. It takes Venus 243 Earth days to spin on its axis, but only 224 Earth days to orbit the Sun!

EARTH

1

Earth formed around 2 billion years after the Sun formed.

2

The Earth orbits the Sun in a part of space called the Cinderella Zone.

3

Earth is not a perfect sphere.

4

The centre of the Earth is made of metal.

5

The atmosphere is a layer of gases that surrounds the Earth.

6

In one year, the Earth travels half way round the Sun.

7

We get seasons because in summer the Earth is nearer the Sun.

8

All the water in Earth's oceans might have been brought here by icy asteroids and comets.

9

Earth is the planet with the most surface water on it.

10

Earth's atmosphere is stopped from floating away into space by gravity.

EARTH

1

FALSE.
The Sun, Earth and the other planets in the Solar System all formed together, around 4.6 billion years ago.

2

FALSE.
Earth orbits in the Goldilocks Zone. This is a band of space where it is not too cold and not too hot, but just the right temperature for liquid water to exist on a planet. Earth orbits at 150 million km from the Sun.

3

TRUE.
The Earth bulges very slightly around the equator at its middle, due to its rotation.

4

TRUE.
The inner core at the centre of the Earth is a solid ball of iron and nickel.

5

TRUE.
The gases are known as air and they help to keep the Earth warm and protect it from harmful radiation.

6

FALSE.
The Earth makes one complete orbit around the Sun every 365 days and 6 hours.

7

FALSE.
Because the Earth is tilted on its axis, the Sun's rays hit the surface at different angles and the amount of daylight changes through the year, both of which affect how warm or cold it is.

8

TRUE.
Scientists don't know for certain how Earth's oceans formed, but one of the most popular ideas is that icy bodies crashed on the young Earth, dumping huge amounts of water over time.

9

TRUE.
70% of the Earth's surface is covered by water. None of the other planets in the Solar System has surface water, though most have some ice.

10

TRUE.
If Earth was smaller, like Mercury, it wouldn't have enough gravity to keep a large atmosphere.

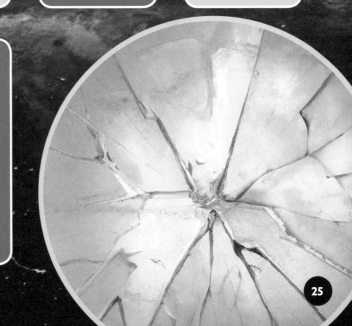

EARTH'S MOON

1 What are the dark patches that you can see on the surface of the Moon at night?

   a. Massive lakes

   b. Solidified pools of lava

   c. Holes in the cheese

2 The Moon formed 4.5 billion years ago – how?

   a. It was a large comet that got captured by Earth's gravity.

   b. A volcano on Venus erupted and ejected a ball of lava that became our Moon.

   c. A planet the size of Mars crashed into Earth and the Moon formed from the debris.

3 How big is the Moon compared to the Earth?

   a. A tenth the size of the Earth

   b. A quarter the size of the Earth

   c. About the same size as the Earth

4 We only ever see one side of the Moon from Earth. True or false?

5 What makes the Moon shine so brightly?

   a. It reflects light from the Sun.

   b. It is covered in glow-worms.

   c. It is surrounded by neon gas.

6 The Moon is moving away from the Earth.
True or false?

7 What important feature of life on Earth does the Moon control?
a. The formation of hurricanes
b. When earthquakes happen
c. The tides of the sea

8 Which day of the week is named after the Moon?
a. Monday
b. Wednesday
c. Sunday

9 An astronaut once hit a golf ball on the Moon.
True or false?

10 What is the Moon made of?
a. Cheese   b. Gas   c. Rock

EARTH'S MOON

1   **b.** The dark patches are known as 'maria' from the Latin for sea, but they are pools of solidified lava.

2   **c.** The collision happened 4.5 billion years ago.

3   **b.** Earth is 12,700 km in diameter, while the Moon measures 3476 km.

4   **True.** The Moon and the Earth are 'tidally locked'. The Moon takes exactly the same time to rotate around its own axis as it does to revolve around its partner, so we only ever see one side of it.

5   **a.** The Moon's surface is quite dark, but it looks bright compared with the surrounding dark space.

6 **True.** The distance between the Moon and Earth is increasing by nearly 4 cm a year!

7 **c.** The Moon's mass pulls on the waters of Earth's oceans, helping to form the tides.

8 **a.** The name came from an Old English word for 'Moon's Day'. In the past, girls born on a Monday were often named Mona.

9 **True.** Alan Shepard was the first person to hit a golf ball on the Moon in February 1971.

10 **c.** The Moon is made of rock, and has lunar soil on the surface.

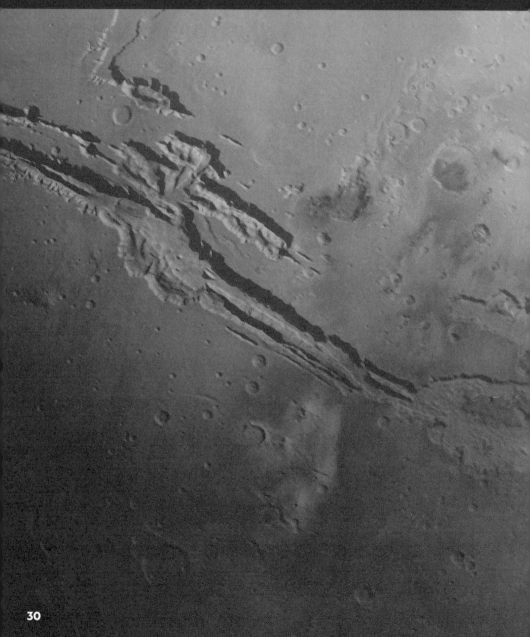

1 The red colour of Mars caused it to be named after which Roman god?
a. The god of war   b. The god of strawberries   c. The god of fire

2 How many moons does Mars have?
a. 0   b. 2   c. 19

3 What is the highest mountain on Mars called?
a. Olympus Mons   b. Olympic Men   c. Mount Mars

4 Pieces of Mars have been found on Earth.
How did they get here?
a. They were brought by Martians.
b. They were in a collision between Mars and Earth.
c. They came from meteorites.

5 NASA's rover Curiosity has been exploring Mars since 2012.
Which of these tools does Curiosity have?
a. A drill to dig into the Martian surface
b. A microscope to examine samples
c. A laser to blast rocks into particles

6 Sunsets on Mars are green.
True or false?

7 How long is a day on Mars?
a. 24.5 Earth hours   b. 24.5 Earth days   c. 24.5 Earth years

8 What liquid flows on Mars?
a. Melted chocolate   b. Water   c. Oil

9 What gives Mars its reddish colour?
a. Rust   b. Ruby crystals   c. Fires

10 Scientists think there is a small possibility that life
might exist on Mars. What sort of life would it be?
a. Primitive fish   b. Lizards   c. Microbes

DID YOU KNOW?
The gravity on Mars is only 37% of that on Earth, which means you could jump around 3 times as high if you were on Mars!

1 **a.** The planet's red colour is noticeable from Earth.

2 **b.** They are called Phobos and Deimos.

3 **a.** Olympus Mons is a Martian volcano standing at over 20 km tall. It is the tallest mountain on any planet in the Solar System.

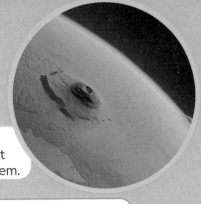

4 **c.** Scientists believe that Mars ejected meteorites which orbited the Solar System before crashing to the ground on Earth.

5 **a, b and c.** Curiosity has many amazing scientific tools, and 17 cameras.

6 **False.** During a Martian sunset, the sky appears blue because the dust in the atmosphere makes it easier for blue light to be seen.

7 **a.** The Martian day is almost the same length as ours.

8 **b.** Scientists believe that about 3.8 billion years ago, Mars had vast amounts of liquid water on its surface, and may have had an ocean that covered one-third of the planet.

9 **a.** Rust is another word for iron oxides, which are plentiful on Mars.

10 **c.** If there is life under the planet's surface or inside rocks, it will be small.

ASTEROIDS

1

There are millions of asteroids in the Solar System and most of them orbit the Sun in a zone just beyond Mars. What is this zone called?

a. The Asteroid Hairband

b. The Asteroid Shoelace

c. The Asteroid Belt

2

Asteroids are orbiting space objects that are too small to be classed as planets.

True or false?

3

What are most asteroids made of?

a. Solid metal

b. Rock and ice

c. Dense gas

4

Vesta is one of the largest asteroids and can be seen with the naked eye when conditions are right. How big is it?

a. About a third of the size of our Moon

b. As big as a house

c. The size of a large car

5

An asteroid hit Earth 65 million years ago— with what result?

a. The impact formed the Pacific Ocean.

b. The event caused the extinction of the dinosaurs.

c. The asteroid bounced off Earth and became the Moon.

6

What shape are asteroids?

a. Square-shaped
b. Star-shaped
c. Any shape

7

What does the name 'asteroid' mean?

a. Big rock
b. Star-like
c. Mini-planet

DID YOU KNOW?

Ceres was the first asteroid ever to be discovered. Giuseppe Piazzi discovered it in 1801.

8

Asteroids are too small to have moons of their own.

True or false?

9

Asteroids often get close to Earth but they don't usually land. Why?

a. Most of them burn up as they enter the atmosphere.

b. They are stopped by a magnetic force field around the planet.

c. They are made of ice which melts before they can land.

10

30 June is International Asteroid Day.

True or false?

ASTEROIDS

1

c.
Most asteroids orbit the Sun in this zone but some cross over with the orbits of Jupiter, Mars and Earth.

2

True.
Asteroids vary in size, but some are as small as a car.

3

b.
Most asteroids are rocky, icy lumps.

4

a.
Vesta is the second-largest asteroid.

5

b.
The catastrophic event caused devastation and climate change that is estimated to have killed 75% of all species.

6

c.
Asteroids are usually irregularly shaped, though some are nearly spherical.

7

b.
Astronomer William Herschel coined the name 'asteroid' in 1802.

8

False.
An asteroid named Ida is only 56 km long, but has a moon called Dactyl.

9

a.
Several groups identify 'near-Earth asteroids' and discover more every year.

10

True.
International Asteroid Day is organised to educate people about asteroids. It happens on 30 June because on that day in 1908 an asteroid hit Siberia causing widespread damage.

DID YOU KNOW?
Although there are 20,000 asteroids close to Earth, there are currently over 990,000 known asteroids in the entire Solar System.

JUPITER

1

Jupiter is named after the king of the Roman gods.

2

Jupiter is easily visible to the naked eye.

3

Humans will never be able to land a spacecraft on Jupiter.

4

Jupiter spins the slowest of all the planets.

5

The first scientist to see that Jupiter has moons was Albert Einstein.

6

Jupiter has 79 moons.

7

One of Jupiter's moons is bigger than the planet Mercury.

8

You could fit 300 Earths into Jupiter.

9

Storms on Jupiter have winds that blow twice as fast as any hurricane winds on Earth.

10

The Great Red Spot, a feature on Jupiter, is actually a storm larger than Earth.

DID YOU KNOW?

Jupiter's magnetic field is around 20 times stronger than Earth's magnetic field, and is the strongest magnetic field in the Solar System.

JUPITER

1

TRUE.
This is fitting as Jupiter weighs more than all the other planets in the Solar System put together.

2

TRUE.
Jupiter is the third-brightest object in the night sky after the Moon and Venus.

3

TRUE.
There is no surface, just an ocean of liquid hydrogen thousands of kilometres deep.

4

FALSE.
It spins the fastest of the planets. One day on Jupiter is only 10 hours long.

5

FALSE.
The first scientist to see that Jupiter has moons was Galileo.

6

TRUE.
Jupiter has 79 confirmed moons. In recent years, 600 more have been discovered but are yet to be officially confirmed.

7

TRUE.
Ganymede is 5300 km in diameter. Mercury is 4900 km in diameter.

8

FALSE.
You could fit 1300 Earths into Jupiter.

9

TRUE.
Storm winds on Jupiter can reach 680 km/h which is more than double the speed of the fastest sustained winds on Earth.

10

TRUE.
The Great Red Spot is not only huge, it has also been blowing for at least 190 years.

SATURN

1 Which of these is NOT true about Saturn?

a. It is the most distant planet that can be seen from Earth with the naked eye.

b. It is a large rocky planet.

c. It has 82 moons.

2 Who is Saturn named after?

a. The Roman god of agriculture and wealth

b. The Norse god of thunder

c. The Greek god of things that have rings around them

3 There is a storm larger than Earth at Saturn's north pole. What shape is this storm?

a. Round

b. Triangular

c. Hexagonal

4 How fast can the wind speeds on Saturn be?

a. As fast as a Formula 1 racing car

b. As fast as a passenger plane

c. Faster than the speed of sound

5 Saturn is the only planet in the Solar System that...

a. Has no seasons

b. Is orbiting the Sun in a different direction

c. Would float on water

6 How many Earths could you fit into Saturn?

a. 763

b. 163

c. 63

7 How long is a day on Saturn?

a. 10.5 Earth hours

b. 10.5 Earth weeks

c. 10.5 Earth years

8 On Saturn, scientists think it could rain...

a. Diamonds

b. Stars

c. Cats and dogs

9 Saturn's shape is a perfect sphere.

True or false?

10 What was the first spacecraft to reach Saturn?

a. *Apollo 11* in 1969

b. *Pioneer 11* in 1979

c. The *International Space Station* in 1988

DID YOU KNOW?

Only four spacecraft from Earth have ever managed to visit Saturn.

SATURN

1

b. Saturn is a gas giant, although it does have a dense metallic core surrounded by rocky material.

2

a. Saturn was considered a god of plenty and peace.

3

c. The sides of the hexagon are each 13,800 km—longer than the diameter of the Earth.

4

c. The winds on Saturn are the second fastest in the Solar System.

5

c. If you could find a bath big enough and dropped Saturn in, the planet would float!

6

a. Saturn is huge compared to Earth.

9 **False.** Saturn is in fact the flattest planet, due to its low density and fast rotation.

7 **a.** Although it is a big planet, Saturn spins very quickly.

10 **b.** *Pioneer 11* launched in 1973 and took 6 years to reach Saturn.

8 **a.** Lightning in its atmosphere turns methane into soot, a form of carbon. As this falls lower and lower, the planet's huge pressure hardens it into diamonds.

SATURN'S RINGS AND MOONS

1

What are Saturn's rings made of?

a. Gold, silver and bronze

b. Roses, violets and daisies

c. Ice, rock and dust

2

What makes Saturn's rings so visible?

a. Reflected sunlight

b. Martians lighting them up with torches

c. Electricity

3

Saturn is the only planet in the Solar System to have a ring system.

True or false?

4

What causes the gaps between the rings?

a. Anti-matter

b. The gravitational pull of small moons nearby

c. Comets smashing through the rings

5

How many separate rings are there?

a. 1

b. 10

c. 1000

6

What is the name of Saturn's largest moon?

a. Timmy

b. Thomas

c. Titan

7

Saturn's largest moon is the only moon in the Solar System to have a thick atmosphere.

True or false?

8

The moon Enceladus has cracks on its surface known as...

a. The tiger stripes

b. Stars and stripes

c. The white stripes

9

How high are some of the mountains on the moon Iapetus?

a. Half as high as Mount Everest

b. The same height as Mount Everest

c. More than twice as high as Mount Everest

10

What is unusual about the moon Mimas?

a. It is heart-shaped.

b. It is banana-shaped.

c. It looks like the Death Star from Star Wars.

SATURN'S RINGS AND MOONS

1

c.
Most of the particles that make up Saturn's rings are very small with a few larger lumps.

2

a.
Saturn's rings are very thin – it is reflected sunlight that makes them so visible.

3

False.
Jupiter, Uranus and Neptune also have rings, but they are much fainter than Saturn's.

4

b.
There are several moons close enough to the rings for their gravity to pull particles out of the rings, leaving a gap.

5

c.
There are three very large rings, four fairly large rings and up to 1000 very small ones.

6

c.
Titan is the second-largest moon in the Solar System after Jupiter's Ganymede.

7

True.
Titan's atmosphere has wind and methane rain, and creates surface features similar to those of Earth.

8

a.
Geysers shoot water out from these cracks (or fissures) to a height of nearly 500 km from the moon's surface.

DID YOU KNOW?
Mimas is the smallest and innermost of Saturn's major moons.

9

c.
The mountain ridge on the moon Iapetus is up to 20 km high.

10

c.
Mimas has a distinctive crater 130 km wide – almost one-third of its diameter.

URANUS

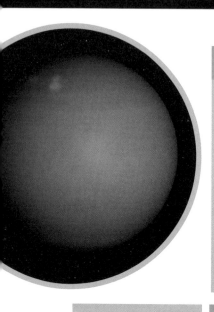

1

What type of planet do scientists describe Uranus as?

a. An ice giant

b. An ice cube

c. An ice cream

2

Uranus is named after which Greek god?

a. The god of the sky

b. The god of ice

c. The god of big planets

3

Which of these facts about Uranus is NOT true?

a. It spins in the opposite direction to most other planets.

b. It lies on its side, with its poles where other planets have their equator.

c. It orbits Jupiter, not the Sun.

4

Uranus is twice as far from the Sun as Saturn.

True or false?

5

How many spacecraft have visited Uranus?

a. 1

b. 4

c. 9

6

How long does it take Uranus to orbit the Sun?

a. 4 Earth years
b. 8 Earth years
c. 84 Earth years

7

How many rings encircle Uranus?

a. 0
b. 13
c. 13,000

DID YOU KNOW?

Uranus is almost four times as big as planet Earth.

8

Uranus can only be seen through a telescope.

True or false?

9

How long is summer on Uranus?

a. 6 Earth months
b. 1 Earth year
c. 21 Earth years

10

It is thought that Uranus's moon Miranda was once smashed into pieces before gradually reassembling itself.

True or false?

URANUS

1

a. The atmosphere of Uranus is mostly hydrogen and helium like the gas giants Saturn and Jupiter. But its mass is made up of 'ices,' water, ammonia and methane. Like Neptune, it is known as an ice giant.

2

a. William Herschel, who discovered the planet, wanted to name it Georgium Sidus, after King George III, but he was over-ruled.

3

c. All the planets orbit the Sun.

4

True. It orbits the Sun at a distance of 2871 million km, twice the distance of Saturn and nearly 20 times the distance from Earth to the Sun.

5

a. NASA's *Voyager 2* spacecraft made the sole visit to Uranus in 1986.

6

c. It takes Uranus 84 years to make one full orbit of the Sun, so in 2033 it will complete only its third orbit since it was discovered in 1781.

7

b. The rings encircling Uranus are very thin and mostly rocks ranging from 10 cm to 2 m across.

8

False. It is just possible to see Uranus with the naked eye, under the right conditions, although no one had identified it as a planet until William Herschel observed it with a telescope in 1781.

9

c. Each of Uranus's four seasons lasts for 21 years.

10

True. The super-rugged moon has massive ridges and craters, and canyons.

NEPTUNE

1 How far away from the Sun is Neptune, compared to Earth?

a. 30 times further

b. 15 times further

c. 5 times further

3 How fast can winds blow on Neptune?

a. As fast as a cheetah

b. As fast as a spitfire plane

c. Twice the speed of sound

2 How was Neptune discovered?

a. By an astronaut in a rocket

b. By someone with extremely good eyesight

c. By mathematical calculation

4 How long does each season last on Neptune?

a. 14 Earth months

b. 41 Earth months

c. 41 Earth years

5 How many moons has Neptune been found to have so far?

a. 2

b. 14

c. 123

6 What is the name of Neptune's largest moon?

a. Tricycle

b. Triton

c. Triangle

7 Neptune is a deep blue colour. What causes this?

a. Oceans of water on the surface

b. Methane gas in the atmosphere

c. Clouds of sapphire crystals

8 In 3.6 billion years' time, Triton will be blown apart by gravity and become part of Neptune's rings.

True or false?

9 What is Neptune's core made of?

a. Solid gold

b. Rock and iron

c. Diamonds and pearls

10 What is Neptune named after?

a. The Roman god of the sea

b. The Norse god of fish

c. The Greek god of swimming pools

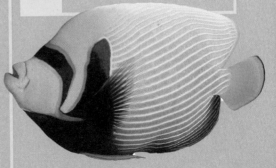

NEPTUNE

1 **a.** Neptune orbits the Sun at an average distance of 4498 million km.

4 **c.** A season on Neptune lasts the equivalent of 41 Earth years.

2 **c.** Astronomers spotted unexpected changes in the orbit of Uranus, and predicted these were caused by the gravity of a planet beyond Uranus. Neptune was finally spotted in 1846, very close to their predicted position.

5 **b.** Neptune's 14th and smallest moon, Hippocamp, was found in 2013 by combining multiple Hubble images.

6 **b.** Triton is unusual in that it orbits Neptune in the opposite direction to the planet's rotation.

3 **c.** Neptune has the fastest winds so far found in the Solar System.

9   **b.** Neptune's core is made of similar material to Earth's core.

7   **b.** The blue methane atmosphere sometimes features darker storms and whitish clouds.

10   **a.** The planet has the astronomical symbol ♆, a stylised version of the god Neptune's trident.

8   **True.** Triton's orbit is getting closer and closer to Neptune. In time, the forces acting on it will cause it to break apart.

PLUTO (A DWARF PLANET)

1 Pluto was discovered by Clyde Tombaugh in 1930 – how old was Clyde at the time?
a. 24   b. 10   c. 72

2 How big is Pluto?
a. The same size as Earth
b. Half the size of Jupiter
c. Two-thirds the size of Earth's moon

DID YOU KNOW?
Although Pluto wasn't discovered until 1930, Percival Lowell predicted its location in 1915 due to inconsistencies that he noticed in the orbits of Uranus and Neptune.

3 Pluto is one-third water. True or false?

4 How many moons does Pluto have?
a. 0   b. 5   c. 10

5 What happened to Pluto in 2006?
a. It was hit by a huge asteroid.
b. The biggest volcano in the Solar System erupted on its surface.
c. It lost its status as a planet.

6 **What is Pluto named after?**
a. The Roman god of the underworld
b. The Greek god of ice and snow
c. The Disney god of dogs

7 **Who gave Pluto its name?**
a. Clyde Tombaugh, who discovered it
b. Venetia Burney, an 11-year-old English schoolgirl
c. Albert Einstein, the physicist

8 **How long is a year on Pluto?**
a. 2 Earth years   b. 248 Earth years   c. 24,800 Earth years

9 **NASA's *New Horizons* spacecraft was launched in 2006 and took over 8 years to reach Pluto.** True or false?

10 **Pluto sometimes swoops inside the orbit of Neptune.** True or false?

PLUTO (A DWARF PLANET)

1 **a.** Tombaugh was a young researcher working for the Lowell Observatory in Arizona, USA, when he made the discovery.

2 **c.** Pluto has a diameter of 2376 km.

3 **True.** Pluto's water is in the form of ice.

4 **b.** The moons are called Charon, Styx, Nix, Kerberos and Hydra.

5 **c.** Pluto is now known as a dwarf planet. There are lots of other objects similar in size to Pluto in the same area of space, and there may be more dwarf planets, too.

6 **a.** The name fits Pluto as it orbits at a great distance from the Sun and so seems dark and unknown.

7 **b.** Venetia was very interested in classical mythology.

8 **b.** Although Pluto's orbit is quite chaotic and can vary between 246 and 249 years.

9 **True.** It had to swing past Jupiter for a gravity boost on the way.

10 **True.** Pluto was last closer to the Sun than Neptune between 1979 and 1999. However, the two planets will never collide with each other.

DWARF PLANETS AND THE OUTER SOLAR SYSTEM

TRUE or FALSE

1

In 1851, there were believed to be 23 planets in our Solar System.

2

There are now exactly 12 recognised dwarf planets.

3

All but one of the known dwarf planets are located in an area called the Kuiper Belt.

DID YOU KNOW?

Pluto and Eris are thought to be the two largest known dwarf planets in the Solar System.

4

There is a dwarf planet called Makemake.

5

Dwarf planets are too small to have moons.

6

Ceres is a dwarf planet that orbits between Mars and Jupiter.

7

One theory says that Neptune's moon Triton was once a dwarf planet.

8

The area where the dwarf planets orbit marks the edge of our Solar System and the start of interstellar space.

9

No spacecraft have yet visited a dwarf planet.

10

Beyond the edge of the Solar System lies a vast shell of tiny, icy objects called the Oort Cloud.

DWARF PLANETS AND THE OUTER SOLAR SYSTEM

TRUE or **FALSE**

1

TRUE.
Astronomers counted lots of very different objects as planets including dwarf planets, asteroids and comets. These objects are now classed differently.

2

FALSE.
No one knows exactly how many dwarf planets there are in the Solar System, and there are frequent new discoveries.

3

TRUE.
The Kuiper Belt is similar to the asteroid belt, but is 20 times wider.

4

TRUE.
Dwarf planets are named after gods from various cultures. Makemake is the Rapanui god of fertility.

5

FALSE.
Several dwarf planets have moons. Haumea has two moons and Pluto has five.

6

TRUE.
Ceres is 952 km in diameter and is the largest object in the asteroid belt.

7

TRUE.
Some scientists think that Triton may have been a dwarf planet which was captured from the Kuiper Belt by Neptune's gravity.

8

TRUE.
There are different definitions of the boundary of the Solar System, but it is commonly thought that the Oort Cloud is the outer edge of the Solar System.

9

FALSE.
Two dwarf planets have been visited by space probes. In 2015, NASA's Dawn mission reached Ceres and its *New Horizons* craft reached Pluto.

10

TRUE.
The Oort Cloud may extend up to 15 trillion km from the Sun.

COMETS

1

What are comets mostly made of?

a. Solid rock

b. Snow and ice

c. Gold and silver

2

Comets are famous for having a large 'tail' streaming out behind them. What is this tail?

a. A stream of ribbons

b. A stream of gas and dust

c. A jet of flames

3

What have comets sometimes been described as?

a. Dirty snowballs

b. Shooting ice

c. Jet pack stars

4

The name 'Comet' comes from the Greek word 'Kometes', which means horse's tail.

True or false?

5

Halley's Comet is the most well-known comet and last appeared in 1986. When will it next appear?

a. 2861

b. Never

c. 2061

6

Comets were created at the same time as the Sun and the planets.

True or false?

7

Which famous historical work of art does Halley's Comet feature in?

a. The Mona Lisa

b. The Bayeux Tapestry

c. The Haywain

8

What happened to Comet Shoemaker-Levy 9 in 1994?

a. It crashed into Jupiter.

b. It passed between the Moon and the Earth.

c. It became inhabited by aliens.

9

In 2014, the *Rosetta* spacecraft became the first to land on a comet. How long had it taken to get there?

a. 10 years

b. 10 months

c. 10 weeks

10

Every August, an astronomical event can been seen in the sky above Earth. What is it called?

a. The Perseids meteor shower

b. A lunar eclipse

c. The Northern Lights

COMETS

1

b.
Comets are made of snow and ice mixed in with dust particles. They may have a rocky core.

2

b.
When the comet gets close to the Sun, solar radiation causes gas and dust to blow out behind the comet.

3

a.
The dusty ice that makes up comets has led to this description.

4

False.
Kometes means 'long hair', which is what the tail on a comet is thought to look like.

5

c.
Halley's Comet reappears every 75–76 years – how old will you be when it next visits in 2061?

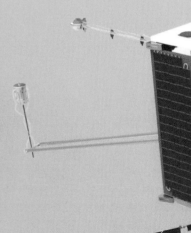

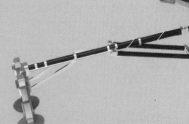

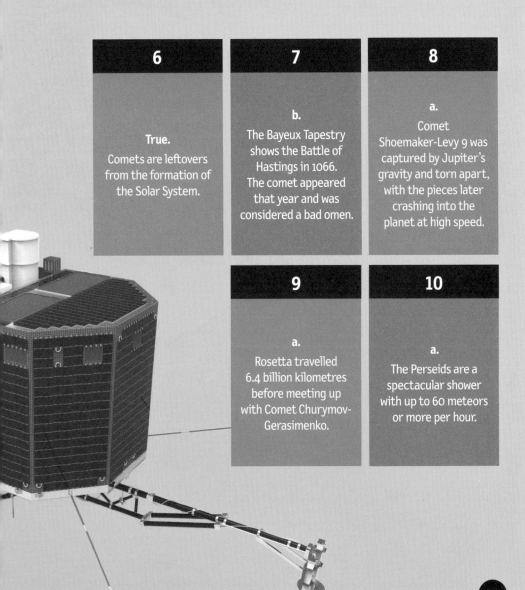

6

True.
Comets are leftovers from the formation of the Solar System.

7

b.
The Bayeux Tapestry shows the Battle of Hastings in 1066. The comet appeared that year and was considered a bad omen.

8

a.
Comet Shoemaker-Levy 9 was captured by Jupiter's gravity and torn apart, with the pieces later crashing into the planet at high speed.

9

a.
Rosetta travelled 6.4 billion kilometres before meeting up with Comet Churymov-Gerasimenko.

10

a.
The Perseids are a spectacular shower with up to 60 meteors or more per hour.

NEBULAE

1 What are nebulae made of?
 a. Clouds of water
 b. Clouds of gas and dust
 c. Clouds of smoke

2 What is the name of the nebula in this picture?
 a. Bull's Eye Nebula
 b. Cat's Eye Nebula
 c. Fish Eye Nebula

3 Which of these is not the name of a real nebula?
 a. Witch's Head Nebula
 b. Monopoly Nebula
 c. Spirograph Nebula

4 What makes the Boomerang Nebula so unusual?
 a. It is reversing backwards to the place it was born.
 b. It is the coldest place in the Universe.
 c. You can only see it from Australia.

5 A nebular cloud the size of Earth would weigh about the same as a cat.
 True or false?

6 Nebulae cannot be seen with the naked eye.
True or false?

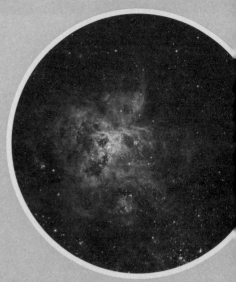

7 The Crab Nebula was discovered in 1840. How did it get its name?

a. The person who discovered it was eating a crab sandwich.

b. It is in the Cancer constellation.

c. It had the same shape as a crab.

8 Which of these is NOT true about nebulae?

a. They are made of water vapour, like clouds.

b. They are formed from dying stars.

c. They are regions where new stars are being made.

9 Which of these is the name of a nebula?

a. Ant Nebula

b. Tarantula Nebula

c. Butterfly Nebula

10 The Sun will form a nebula when it reaches the end of its life.
True or false?

NEBULAE

1

b.
Nebulae are mostly made of hydrogen, helium and dust particles.

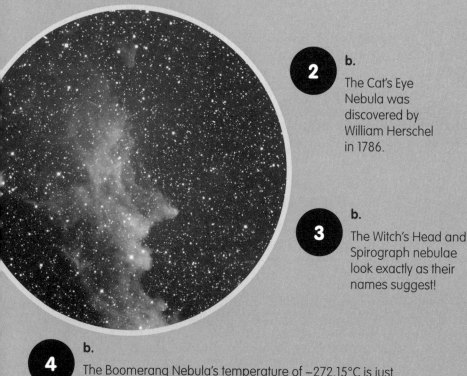

2

b.
The Cat's Eye Nebula was discovered by William Herschel in 1786.

3

b.
The Witch's Head and Spirograph nebulae look exactly as their names suggest!

4

b.
The Boomerang Nebula's temperature of −272.15°C is just one degree above absolute zero, which is −273.15°C.

5

True.
The material in a nebula is very spread out. Nebulae are denser than the space around them, but less dense than any vacuum that we have ever created on Earth.

6

False.
You can see several nebulae, including the Orion Nebula. This is the middle 'star' in the sword of the constellation of Orion.

7

c.
William Parsons, the third Earl of Rosse, first saw it through a telescope. The picture he drew of it looked like a crab.

8

a.
'Nebula' is from the Latin word for cloud, but unlike the clouds we see in the sky, they are not made of water vapour.

9

a, b and c.
Nebulae are often named after the creature or object whose shape they resemble.

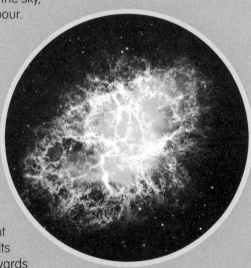

10

True.
The Sun will become a red giant star before running out of fuel. Its outer layers will then blow outwards forming a vast nebula with a white dwarf star at its core.

STARS

1

A star forms in a huge cloud of gas and dust called a...
a. Spatula
b. Nebula
c. Nucleus

2

How does a star form?
a. An area of gas and dust collapses under its own gravity and begins heating up.
b. Interstellar winds spin the cloud into a ball.
c. The gas catches fire.

3

What is the colour of very hot stars?
a. Blue
b. Red
c. Yellow

4

Our Sun is a star.
True or false?

5

Which of these is NOT a type of star?

a. White dwarf

b. Blue giant

c. Yellow titan

6

Giant stars live longer than smaller stars.

True or false?

7

Some stars are destroyed in a massive explosion called a...

a. Starburst

b. Firework

c. Supernova

8

What is the brightest star in the night sky?

a. Canopus

b. Sirius

c. Venus

9

A shooting star is not a star at all. What is it really?

a. Fragments of an ancient supernova

b. Tiny bits of rock burning up in Earth's atmosphere

c. Stray comets

10

Why do stars appear to twinkle?

a. The Earth's atmosphere disturbs the path of their light rays.

b. They flash as a warning to other stars.

c. Giant space creatures keep flying in front of them.

STARS

1

b.
Nebula is the Latin word for cloud or fog.

2

a.
The gas heats up and becomes a 'protostar', a hot core within the collapsing cloud that will one day become a star.

3

a.
Blue stars are very hot, yellow are less hot and red are the least hot stars.

4

True.
The Sun is the closest star to Earth.

5

c.
Other star types include red dwarfs, red giants and brown dwarfs.

6

False.
Giant stars are brighter and so they burn out quicker than smaller, less-bright stars.

7

c.
Supernovae are the biggest explosions in the universe. Some have been visible during the day on Earth.

8

b.
Sirius is 8.6 light-years away, but if it replaced our Sun it would shine 20 times as brightly.

DID YOU KNOW?

The Sun is planet Earth's closest star.

9

b.
The tiny bits of dust and rock are called meteoroids. If any pieces survive hitting the atmosphere and fall to Earth they are called meteorites.

10

a.
The twinkling is just an illusion caused by turbulence in the Earth's atmosphere.

GALAXIES

1 **What is a galaxy?**
a. A vast cloud of gas and dust
b. A large group of stars, bound together by gravity
c. The scattered remains of a star

2 **Galaxies can have hundreds of millions of stars.** True or false?

3 **What is the name of the galaxy that our Sun is in?**
a. The Swirly Whirly
b. The Milky Way
c. The Chocolate Bar

4 **Galaxies are classified by their shapes. Which of these is NOT a type of galaxy?**
a. Spiral   b. Elliptical   c. Corkscrew

5 **Which if these is NOT the real name of a galaxy?**
a. The Sheepdog Galaxy
b. The Whirlpool Galaxy
c. The Sombrero Galaxy

6 What is the name of this galaxy?
a. The Big Wheel Galaxy
b. The Pinwheel Galaxy
c. The Catherine Wheel Galaxy

7 What is the name of the galaxy that is closest to ours?
a. Andromeda
b. Andrea
c. Angela

8 What lies at the centre of our galaxy?
a. A star 100 times bigger than the Sun
b. A black hole
c. Nothing, just empty space

9 It isn't possible to see other galaxies with the naked eye.
True or false?

10 What is going to happen to the Milky Way and Andromeda in about 4 billion years?
a. They are both going to double in size.
b. They are going to collide.
c. They are going to explode.

GALAXIES

1 **b.** The stars may be billions of kilometres apart but they are still part of the same galaxy.

2 **True.** Although some dwarf galaxies may only have 1000 stars and larger galaxies may have billions.

3 **b.** The name describes the white cloudy band that you can see in the night sky.

4 **c.** The three main types of galaxy are spiral, elliptical and irregular.

DID YOU KNOW?

In 1926, Edwin Hubble set up a system to classify galaxies into four types: spiral, lenticular, elliptical and irregular galaxies.

5 **a.** The Sombrero and Whirlpool galaxies are named after their appearance.

6   **b.** The Pinwheel Galaxy (also known as Messier 101) is a spiral galaxy.

7   **a.** Andromeda is the same age and size as the Milky Way, but contains many more stars.

8   **b.** The black hole is known as Sagittarius A* and it has a mass of around 4.1–4.5 million times the mass of the Sun.

9   **False.** It is possible to see Andromeda, which is the furthest object in space that can be seen with the naked eye.

10   **b.** Scientists originally thought that this collision would happen sooner, which means that the Milky Way will survive in its current form for a bit longer than first predicted.

THE UNIVERSE

1

Astronomers think that the Universe started with what?

a. The Big Bang

b. The Enormous Explosion

c. The Big Chill

2

How old is the Universe?

a. 1.3 billion years

b. 13.8 billion years

c. 138 billion years

3

What is the geometric shape of the Universe thought to be?

a. Round

b. Square

c. Flat

4

Where were the atoms that make up our bodies formed?

a. In a science laboratory

b. Underground

c. Inside exploding stars

5

The Universe is currently expanding. True or false?

6

The 'observable Universe' is the part that...

a. Has been explored by spacecraft

b. Is lit up by the Sun's rays

c. We can see from Earth

7

If a star is 5000 light-years away, then we are seeing it...

a. As it was 5000 years ago

b. As it is now

c. As it will be in 5000 years time

8

There are more stars in the Universe than there are grains of sand on all the beaches on planet Earth.

True or false?

9

Most of the Universe is made up of what scientists call 'dark energy'. What is this?

a. Nobody really knows

b. A magnetic force field

c. The energy from a black hole

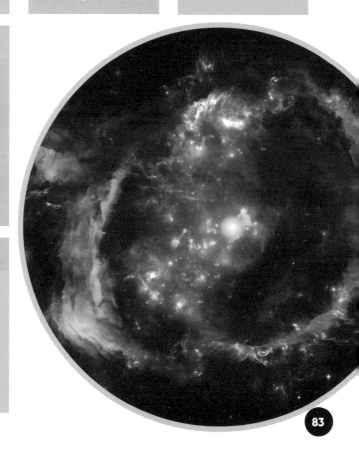

10

What is the name of the theory of how the Universe will end?

a. The Big Rip

b. The Big Freeze

c. The Big Crunch

THE UNIVERSE

1

a.
The Big Bang describes how the Universe suddenly expanded from being extremely dense and hot into the huge cosmos we can observe today.

2

b.
Our Solar System formed 4.6 billion years ago.

3

c.
Scientists have worked this out by looking at the pull of gravity and the rate of expansion.

4

c.
Our world is built from elements that were formed in supernovae—we are all made of stardust!

5

True.
The Universe is expanding and individual galaxies are moving away from each other.

6

c.
This also includes what we can see on Earth from telescopes that are out in space.

7

a.
Even at its very fast speed, that light from the star set off 5000 years ago. So looking at the night sky is like looking back in a time machine!

8

True.
There may be around ten times the number of stars as there are grains of sand on all the beaches on Earth.

9

a.
Astronomers have worked out that most of the Universe is made up of 'dark matter' (27%) and 'dark energy' (68%). No one knows what dark matter and dark energy really are.

10

a, b and c.
No one is certain how the Universe will end, but there are some very different theories about it.

BLACK HOLES AND OTHER STRANGE OBJECTS

TRUE or **FALSE**

1

Black holes are just a theory – they have never been observed.

2

A black hole is an area of space where gravity is so strong that not even light can escape from it.

3

Some supermassive black holes are much heavier than the Sun.

4

A teaspoon of material from a neutron star would weigh as much as a mountain.

5

Despite its name, a magnetar is a type of neutron star that has no magnetic field.

6

A single quasar can be thousands of times brighter than a galaxy such as the Milky Way.

7

There is a fountain of anti-matter in the middle of our galaxy.

8

A zombie star is created when a supernova doesn't fully explode.

9

A pulsar is a star that beats in and out like a heart.

10

Centaurus A is a galaxy that ate another galaxy.

BLACK HOLES AND OTHER STRANGE OBJECTS

TRUE or **FALSE**

1

FALSE.
An image of a black hole was taken in 2019.

2

TRUE.
Black holes form when massive stars collapse at the end of their life.

3

TRUE.
Supermassive black holes are the largest type of black hole.

4

TRUE.
Neutron stars are veryu dense, having the mass of two or more of our Suns packed into a sphere 20 km cross. That teaspoon of material would weigh 1 billion tonnes.

5

FALSE.
Magnetars are the most powerful magnetic stars in the Universe.

6

TRUE.
A quasar is caused by gas emitting energy as it falls into a black hole.

7

TRUE.
The fountain is spurting out 15 billion tonnes of anti-matter every second into a plume that is 3500 light-years long.

8

TRUE.
Some types of supernova leave a remnant star behind after the explosion – a zombie star.

9

FALSE.
A pulsar is a rapidly spinning neutron star that emits beams of particles that sweep across the sky. It is more like a lighthouse than a heart.

10

TRUE.
Centaurus A has a dark disc across its middle – the remnants of a spiral galaxy that it gobbled up.

ASTRONOMY

1

Astronomers study the positions of the stars to predict a person's future.

True or false?

2

Galileo was one of the first astronomers to use a telescope. In 1610, he became the first human ever to observe...

a. Venus

b. Jupiter's moons

c. The far side of the Moon

3

Ptolemy was one of the earliest astronomers. What did he believe was at the centre of the Universe?

a. A black hole

b. Earth

c. The Sun

4

Which of these is NOT a type of telescope?

a. Wifi telescope

b. Radio telescope

c. Optical telescope

5

What type of telescope is this?

a. Optical telescope

b. Pirate's telescope

c. Radio telescope

6

Which of these space objects can you see with ordinary binoculars?

a. Four of Jupiter's moons

b. The planet Neptune

c. The Andromeda spiral galaxy

7

What important discovery did astronomer Edwin Hubble make in 1924?

a. The Milky Way is not the only galaxy in the Universe.

b. Earth is spinning towards a black hole.

c. There are aliens on Mars.

8

The Hubble Space Telescope is named after Edwin Hubble. But why was it put in space?

a. To be outside Earth's atmosphere.

b. To be closer to the stars.

c. To get free solar energy from the Sun.

9

Which well-known British astronomer used to be a pop star?

a. Professor Green

b. Dr Dre

c. Professor Brian Cox

10

How did the 17th century physicist and astronomer Isaac Newton come up with his theory of gravity?

a. By falling off a horse

b. By watching an apple fall from a tree

c. By going into space

ASTRONOMY

1

False.

It is astrologers who read the stars to predict someone's future. Astronomers are scientists who study space and all the objects in it.

2

b.

Galileo was also the first person to see Saturn's rings, although he thought they were moons.

3

b.

The astronomer believed that the Sun, planets and stars all orbited the Earth. This theory was described as 'geocentric' because geo was Greek for Earth.

4

a.

Radio telescopes receive and amplify radio waves from space objects. Optical telescopes magnify light waves by using either lenses to refract (bend) light, or mirrors to reflect it.

5

c.

This is the radio telescope at Jodrell Bank in England. It gathers information from space objects that emit radio waves.

6

a, b and c.
Binoculars are a really easy and fun way to start observing amazing things in the night sky.

7

a.
Hubble made the very important discovery that there are galaxies beyond our own.

8

a.
Our atmosphere blocks and scatters some light from space, so Hubble was put in space to get a clearer view.

9

c.
In the 1990s, Professor Brian Cox played keyboards in a band called D:Ream.

10

b.
Seeing an apple fall from a tree when he was sitting in a garden helped Isaac Newton to understand and explain how gravity works.

DID YOU KNOW?
The James Webb Space Telescope will weigh about the same as a small bus! It will succeed the Hubble Telescope.

THE SPACE RACE

1

What was the first living thing sent into space?

a. Neil Armstrong, a human

b. A frog called Spacehopper

c. Fruit flies

2

What was the Space Race?

a. A competition between the USA and the Soviet Union to be the first country to go into space

b. A challenge to see who could build the fastest spacecraft

c. A 100 m dash by two astronauts on the Moon

3

This is *Sputnik 1*, which became the first-ever artificial satellite when it was launched by the Soviet Union in 1957. Why is it so shiny?

a. To make it easier to spot with a telescope

b. To reflect heat

c. To scare away aliens

4

How big was *Sputnik 1*?

a. The size of a beach ball

b. The size of an orange

c. The size of a hot air balloon

5

The first US satellite, *Explorer 1*, launched in 1958. What did it study?

a. Cosmic rays

b. Stingrays

c. X-rays

Alexei Leonov made the first spacewalk in 1965. What problem did he encounter?

8

a. His spacesuit swelled up and he couldn't fit back in his spacecraft.

b. The cable attaching him to the spacecraft broke and he floated away.

c. He was hit by a meteoroid.

Yuri Gagarin was the first human to travel in space. How did he get back to Earth?

6

a. He got a lift from an alien.

b. He landed his rocket in a car park.

c. He ejected from his spacecraft and parachuted to Earth.

What was the name of the first woman to go into space?

9

a. Valentina Tereshkova

b. Venus Williams

c. Halle Berry

There has never been a rocket launched into orbit from the UK. True or false?

7

What kind of animal was Laika, the first animal to orbit Earth?

10

a. Pig   b. Rat   c. Dog

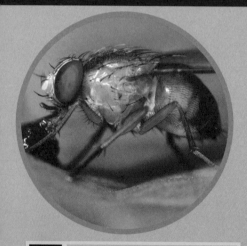

b.

3

The shiny surface of *Sputnik 1* protected the contents inside from the heat.

a.

4

At the time, the technology wasn't available for launching anything bigger into space.

c.

1

The first living creatures sent to space were fruit flies in 1947.

a.

2

The Space Race began around 1955 during the Cold War—a period of great tension between the Soviet Union and the USA.

a.

5

Cosmic rays are high-energy particles that zoom through space.

6

c.
The *Vostok 1* spacecraft had no way of landing safely, so the cosmonaut had to eject from it as it approached Earth.

7

True.
The race is on!

8

a.
Alexei Leonov's spacesuit was too hot inside—he had to drain out air from it before he could get back inside his spacecraft.

9

a.
Russian Valentina Tereshkova went into space in 1963, and is still the only woman to have flown solo in space.

10

c.
Laika was a stray dog from Moscow, Russia, who went into space in *Sputnik 2*.

THE MOON LANDINGS

1 Which Apollo mission was the first to land on the Moon?
a. Apollo 13
b. Apollo 11
c. Apollo 1

2 What did the astronauts on that mission leave behind on the Moon?
a. An American flag
b. Their camera
c. Bags of rubbish

3 What was the name of the lunar module that the astronauts used to land on the Moon?
a. Sparrow
b. Budgie
c. Eagle

4 How long did it take Apollo 11 to travel to the Moon?
a. 3 hours   b. 3 days   c. 3 months

5 The first human to step onto the Moon's surface was Neil Armstrong.
True or false?

6

What went wrong with the Apollo 13 mission?

a. The astronauts ran out of food.

b. An oxygen tank exploded on the way to the Moon.

c. The astronauts were left stranded on the Moon.

7

The last three Apollo missions took 4-wheeled lunar roving vehicles with them to travel around the Moon's surface. What were they known as?

a. Dune buggies   b. Moon buggies   c. Moon tanks

8

How many people have walked on the Moon?

a. 12

b. 2

c. 99

9

Who was the second man to walk on the Moon?

a. Buzz Aldrin

b. Buzz Lightyear

c. Michael Collins

10

The Soviet Union sent a successful mission to the Moon soon after the last Apollo mission in 1972. True or false?

THE MOON LANDINGS

1

b.
There were several test missions and cancelled missions before Apollo 11 was successful.

2

a, b and c.
In order to keep the weight of the lunar module as low as possible, the astronauts left behind anything they no longer needed.

3

c.
Eagle was carried on the command module Columbia, before separating off to land on the Moon.

4

b.
Apollo 11 took off on 16 July and reached the Moon's orbit on 19 July. The Moon landing was on 20 July.

5

True.
Neil Armstrong was the first human to step on any celestial body other than Earth.

6

b.

The astronauts had to cancel their Moon landing, but they adapted their spacecraft, looped round the Moon and made it safely back to Earth.

7

b.

Each moon buggy drove around 30 km on its mission. They are still up there on the Moon!

8

a.

Of the 24 astronauts who have travelled to the Moon, 12 have taken steps on it.

9

a.

Buzz Aldrin was the second man to walk on the Moon. Michael Collins stayed in orbit in the command module.

10

False.

No nation other than the United States has landed a human on the Moon.

THE SPACE SHUTTLE

1

The Space Shuttle was the world's first reusable spacecraft.

True or false?

2

When did the Space Shuttle first fly?

a. 1881

b. 1981

c. 2018

DID YOU KNOW?

If you combine the distance travelled of all five of the Space Shuttle orbiters, it comes to almost 1.3 times the distance between Earth and Jupiter – nearly 827 million kilometres.

3

Which of these is NOT the name of a Space Shuttle orbiter?

a. Discovery

b. Columbia

c. Pluto

4

How many missions did the Space Shuttle make into space?

a. 13

b. 78

c. 135

5

The Space Shuttle is the largest spacecraft ever launched into orbit.

True or false?

6

The Space Shuttle is covered with special tiles — why?

a. To trap solar energy while in orbit.

b. To camouflage it.

c. To protect it from heat.

7

What happened to the two solid rocket boosters after they had used up their fuel?

a. They drifted back down to Earth on parachutes.

b. They were ejected into space.

c. They burned up on re-entry into the atmosphere.

8

The Space Shuttle once took ants into space. True or false?

9

While in orbit, the Space Shuttle travelled around Earth at a very high speed. What did this mean for the crew?

a. It made them travel sick.

b. It gave them jet lag.

c. They saw a sunrise or sunset every 45 minutes.

10

In 2009, what did astronaut Michael J. Massimino become the first person to do?

a. Post a Facebook status in space

b. Send a tweet from space

c. Take a selfie using the Hubble space telescope

THE SPACE SHUTTLE

1

True.
It took off like a rocket, then jettisoned its fuel tanks and later returned to Earth like a glider.

2

b.
The Space Shuttle's first flight was originally planned in 1968 before humans had gone to the Moon.

3

c.
There were 5 shuttles: Columbia, Atlantis, Challenger, Discovery and Endeavour.

4

c.
The last of the 135 missions was in 2011.

5

True.
The shuttle is the size of a small jet airliner.

6

c.
During re-entry to Earth's atmosphere, the outside of the orbiter heats up to more than 1500°C.

7

a.
The solid rocket boosters drifted back down to Earth and were reused.

8

True.
The Space Shuttle took ants into space in 2003.

9

c.
The Space Shuttle crew could see the shadow of night falling across the Earth.

10

b.
Michael J. Massimino tweeted: 'From orbit: Launch was awesome!! I am feeling great, working hard, & enjoying the magnificent views, the adventure of a lifetime has begun!'

FUTURE SPACE MISSIONS

1

NASA's SWOT satellite will be searching space for flies.

2

The Martian Moons eXploration is a mission to explore the two moons of Mars.

3

A spacecraft will bring pieces of an asteroid back to Earth in 2023.

4

The NASA plan to land the first woman on the Moon by 2024 is called 'Apollo'.

5

In 2025, the largest telescope on Earth will open. It will be called 'The Extremely Large Telescope'.

6

The European Space Agency's JUICE probe will take 10 years to reach its final destination.

7

NASA is planning to send humans to Mars by 2027.

8

The first rockets to be launched from the UK will be taking off from Scotland.

9

The *Psyche* mission aims to visit a huge ball of metal in the asteroid belt.

10

There are plans to turn the International Space Station into a luxury hotel.

FUTURE SPACE MISSIONS

1

FALSE.
The Surface Water and Ocean Topography satellite will be monitoring the water on Earth.

2

TRUE.
The mission aims to explore Phobos and Deimos.

3

TRUE.
The OSIRIS-REx mission set off in 2016 to visit the asteroid Bennu, and will bring back a 60 g sample of its surface.

4

FALSE.
The plan to land the first woman on the Moon by 2024 is called Artemis, after the Greek goddess who was associated with the Moon.

5

TRUE.
'The Extremely Large Telescope' (ELT) is a very accurate name – the ELT will have a mirror 39 m in diameter and be 13 times more powerful than the current largest Earth telescope.

6

TRUE.
JUICE launches in 2022 and will reach Jupiter in 2029, but is scheduled to orbit Jupiter's moon Ganymede between 2032 and 2033.

7

FALSE.
The mission to send humans to Mars is being planned, but it won't be until the 2030s.

8

TRUE.
The UK's first rocket launch site has been approved in the far north of Scotland.

9

TRUE.
Astronomers think the asteroid is made of iron and nickel and is the core of a protoplanet that got smashed apart.

10

TRUE.
The Russian space agency has drawn up plans to add sleeping quarters. A one-week trip would cost you $40 million.

THE INTERNATIONAL SPACE STATION

1

Why was the International Space Station (ISS) built in space and not on Earth?

a. It was made from recycled space junk.

b. It was too big to launch from one place.

c. So that solar-powered tools could be used.

2

It is not possible to see the ISS with the naked eye from Earth.

True or false?

3

How many times does the ISS orbit the Earth in 24 hours?

a. 15 times

b. 3 times

c. once

4

How long has the ISS been continuously occupied for (as of November 2020)?

a. 5 years

b. 20 years

c. 50 years

5

How many different people have visited the ISS?

a. 2

b. 790

c. 240

6

What did astronaut Chris Hadfield do in the ISS in 2013 that caught the world's attention?

a. Played the song 'Space Oddity' on guitar and sang along

b. Crashed it into the Moon

c. Became the first person to juggle 9 balls

7

Astronauts spend time running on treadmills while on the ISS – why?

a. The treadmills generate energy that the ISS can use.

b. To keep strong and fit.

c. To stop themselves from getting bored.

8

The ISS is also a science lab. Which of these discoveries was made on board?

a. Some bacteria can survive in space.

b. Flames from a candle form a sphere in space.

c. Crystals can grow to huge sizes when there is no gravity.

9

Water vapour from astronauts' breath is recycled on the ISS and used for what?

a. Washing the solar panels

b. Boiling pasta

c. Drinking water

10

Which British astronaut visited the ISS in 2015–16?

a. Tim Peake

b. Tim Henman

c. Jim Henson

THE INTERNATIONAL SPACE STATION

1

b. The ISS has been put together piece by piece. It is now larger than a football pitch and weighs 420 tonnes.

2

False.
The ISS is the third-brightest object in the sky and is often easy to spot.

3

a. The ISS circles the Earth every 93 minutes.

4

b. The first residents arrived on 2 November 2000 and the ISS has been continuously inhabited since then.

5

c. There have been 151 American visitors, 49 Russian visitors and representatives from 17 other countries.

6

a. Chris Hadfield (left) performed (guitar and singing) and broadcast the video. 'Space Oddity' was written in 1969 by singer David Bowie about an astronaut on a mission that goes wrong.

7

b. Being weightless in space causes muscles to weaken and also bone loss. Astronauts exercise to stay as strong and fit as they can.

8

a, b and c. The unique environment of the ISS has led to some amazing scientific discoveries.

9

c. Resources are very precious in space and as much is reused as possible.

10

a. Tim Peake (right) spent 6 months in space.

EXTRATERRESTRIAL LIFE

1 We know for a fact that life exists on planets beyond our Solar System.

True or false?

2 What are exoplanets?

a. Planets where eggs have been found

b. Planets where X-Men live

c. Planets that orbit stars other than our Sun

3 How many exoplanets have been discovered so far?

a. 4292

b. 42

c. 4

4 What did astronomers once think existed on Mars?

a. Water slides

b. Canals

c. Fountains

5 Scientists think they may have found evidence of life in the atmosphere around Venus.

True or false?

6 Research for life beyond Earth is known as SETI. What does SETI stand for?

a. Search for Extraterrestrial Intelligence

b. Studying E.T. Intently

c. Scanning Extremely Tiny Images

7 In 1977, a radio telescope detected a transmission that many people think could be from intelligent life. What is the transmission known as?

a. The Skywalker Series

b. The MegaBurst

c. The Wow! Signal

8 What is the name of the biggest project currently searching for extraterrestrial communications?

a. Search for a Star

b. Breakthrough Listen

c. Radio Gaga

9 When a star began dimming strangely in 2011, scientists thought there could be an alien megastructure being built around it. True or false?

10 Where will the *Dragonfly* spacecraft be going in search of conditions that could support life?

a. The Sun

b. Titan, Saturn's largest moon

c. The planet Krypton

EXTRATERRESTRIAL LIFE

1

False.
We don't have concrete evidence of extraterrestrial life, but scientists are looking for that evidence in many ways.

2

c.
The Kepler space telescope spent 9 years searching space for exoplanets until it ran out of fuel in 2018.

3

a.
The 4292 exoplanets lie in 3223 star systems —714 of these systems have more than one planet.

4

b.
Astronomers in the 19th century looked at Mars and thought they saw a network of long straight canals. With our more powerful telescopes we know this was just an optical illusion.

5

True.
Although the surface of Venus is extremely hot and inhospitable, the atmosphere is similar to Earth's and may contain tiny microbes.

6 **a.**
International SETI programmes have been going on since the 1960s.

9 **True.**
A huge alien structure was a real possibility for the strange observation. It is now thought to be caused by huge plumes of dust.

7 **c.**
The Wow! Signal was so unusual that the astronomer who spotted it wrote 'Wow!' on his computer printout—hence the name.

10 **b.**
Titan has a complex chemistry on its surface and an interior water ocean—it could be a place where life might evolve.

8 **b.**
Launched in 2016, the project is searching stars and galaxies for radio signals and laser transmissions.

DID YOU KNOW?
While there is no concrete proof that once upon a time there was life on Mars, its surface environment had water that could have been habitable for microorganisms.

CONSTELLATIONS

1 A constellation is a group of stars that forms a pattern. True or false?

2 What is the constellation Ursa Major also known as?
a. The Great Dog
b. The Little Lion
c. The Great Bear

DID YOU KNOW?
Crux is the smallest constellation. It covers less than 0.2% of the sky.

3 What constellation is said to resemble a pair of twins?
a. Libra   b. Gemini   c. Sagittarius

4 Hydra is the largest constellation. What is it named after?
a. A monstrous serpent
b. A one-eyed giant
c. An evil fish

5 Farmers used to use the constellations to help them decide when to plant and harvest crops. True or false?

6 Which of these countries does NOT have a constellation on its flag?

a. New Zealand   b. South Africa   c. Australia

7 The stars in each constellation are all quite close together. True or false?

8 One of the most recognisable patterns of stars is The Plough. What shape does it make?

a. A saucepan

b. A spade

c. A tractor

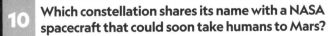

9 All of the constellations are within the Milky Way. True or false?

10 Which constellation shares its name with a NASA spacecraft that could soon take humans to Mars?

a. Aries   b. Orion   c. Scorpio

CONSTELLATIONS

1   **True.** There are 88 recognised constellations forming a variety of patterns, from mythical creatures to tools.

2   **c.** Ursa Major – or the Great Bear – is the third-largest constellation.

3   **b.** The bright stars at the figureheads of Gemini are named after the mythical twins Castor and Pollux.

4   **a.** Hydra is named after a monstrous serpent that had many snake heads on a dog-like body.

5   **True.** The changing positions of the constellations acted like a calendar.

DID YOU KNOW?

Hydra is the biggest constellation of stars by area. It takes up over 3% of the night sky.

6 **b.** Australia and New Zealand both feature the Southern Cross constellation on their flags.

7 **False.** The stars in constellations seem close from our point of view but they can be different brightnesses and distances away in deep space. In the constellation Cygnus, the swan, the faintest star is the closest and the brightest star is the furthest away!

8 **a.** The Plough is a small pattern of stars known as an asterism. It forms part of the larger Ursa Major constellation.

9 **True.** The constellations are all close enough to be seen by the naked eye.

10 **b.** The Orion spacecraft is designed to carry 4 astronauts and it is planned to head to Mars in the 2030s.

SPACE IN OUR CULTURE

1. **If a spacecraft in *Star Trek* is travelling at 'Warp 2', how fast is it going?**
a. As fast as a milk float
b. Twice the speed of sound
c. Eight times the speed of light

2. **The Doctor in *Doctor Who* comes from which planet?**
a. Tatooine   b. Neptune   c. Gallifrey

3. **What job did Wall-E the robot have?**
a. Serving food in a restaurant
b. Building hover-bikes
c. Compacting and stacking garbage

4. **Who wrote the book *The War of the Worlds*?**
a. J.K. Rowling
b. Steven Spielberg
c. H.G. Wells

5. **What do the letters E.T. stand for in the film about the lost alien?**
a. Escaping Triton
b. Extra Terrestrial
c. Extending Time

 6 What nickname does Peter Quill give himself in *Guardians of the Galaxy*?

a. Star Captain b. Star King c. Star Lord

 7 Which movie was set 'A long time ago in a galaxy far, far away...'?

a. *Star Trek*

b. *Star Wars*

c. *Battlestar Galactica*

8 The Disney character Pluto shares his name with the dwarf planet. But what kind of creature is Pluto?

a. A cat b. A dog c. A mouse

 9 What, according to *The Hitchhiker's Guide to the Galaxy*, is the answer to the Ultimate Question of Life, the Universe and Everything?

a. Be excellent to each other.

b. May the Force be with you.

c. 42.

 10 'I was born on the planet Krypton and called Kal-El. My planet was destroyed and I was sent to Earth as a baby where I was taken in by the Kent family. Who am I?'

a. Batman   b. Superman   c. Iron Man

SPACE IN OUR CULTURE

1 **c.** A warp drive can propel a spacecraft faster than light-speed.

2 **c.** Gallifrey is the home of the Time Lords.

3 **c.** Wall-E was left behind on Earth to stack up the rubbish.

4 **c.** H.G. Wells wrote several famous science-fiction books, including *The War of the Worlds*, *The Time Machine* and *The Invisible Man*.

5 **b.** Extra Terrestrial means 'beyond Earth'.

THE H.G. WELLS SOCIETY

H.G. WELLS
AUTHOR
1866~1946
LIVED
AND WORKED HERE
1930~1936

6 **c.** Not many of the other characters in *Guardians of the Galaxy* call Peter Quill 'Star Lord', though...

7 **b.** The line has appeared at the start of all the main Star Wars films since the first film in 1977.

8 **b.** Pluto is Mickey Mouse's pet.

9 **c.** The answer to the Ultimate Question of Life, the Universe and Everything was worked out by the computer Deep Thought over 7½ million years.

10 **b.** Superman the character first appeared in a comic book in 1938.

TIE BREAKER

If there's a draw between two players, try this
tie breaker question... closest answer wins!

THE TALLEST MOUNTAIN KNOWN TO HUMANS IS ON THE ASTEROID, VESTA. HOW TALL IS IT IN KILOMETRES?

The mountain on Vesta is a staggering 22 km tall.

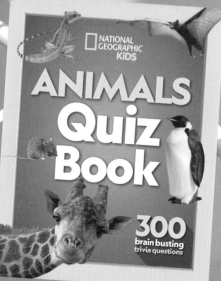

NATIONAL GEOGRAPHIC KIDS

Quiz books

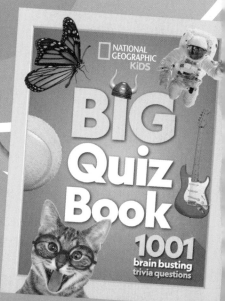